Farm Animals

GOATS ON THE FARM

By Rose Carraway

Gareth Stevens
Publishing

Please visit our website, www.garethstevens.com. For a free color catalog of all our high-quality books, call toll free 1-800-542-2595 or fax 1-877-542-2596.

Library of Congress Cataloging-in-Publication Data

Carraway, Rose.
Goats on the farm / Rose Carraway.
    p. cm. — (Farm animals)
Includes index.
ISBN 978-1-4339-7353-6 (pbk.)
ISBN 978-1-4339-7354-3 (6-pack)
ISBN 978-1-4339-7352-9 (library binding)
1. Goats—Juvenile literature. I. Title.
SF383.35.C37 2013
636.3'9—dc23

                                        2012003050

First Edition

Published in 2013 by
Gareth Stevens Publishing
111 East 14th Street, Suite 349
New York, NY 10003

Editor: Katie Kawa
Designer: Andrea Davison-Bartolotta

Photo credits: Cover, p. 1 Hemera/Thinkstock; p. 5 Dmitrijs Bindemanis/Shutterstock.com; p. 7 Angela Luchianiuc/Shutterstock.com; p. 9 Halina Yakushevich/Shutterstock.com; pp. 11, 24 (beard) Judy Kennamer/Shutterstock.com; p. 13 Natalia Kashina/Shutterstock.com; pp. 15, 24 (wool) sherpa/Shutterstock.com; p. 17 TOMO/Shutterstock.com; p. 17 (inset) Angelo Gilardelli/Shutterstock.com; p. 19 Ignite Lab/Shutterstock.com; p. 21 © iStockphoto.com/Carol Thacker; pp. 23, 24 (cheese) Cultura/Zero Creatives/Getty Images.

Printed in the United States of America

CPSIA compliance information: Batch #CS12GS: For further information contact Gareth Stevens, New York, New York at 1-800-542-2595.

Contents

Goats are smart animals.

Goats like
to stay together.
A group of goats
is called a herd.

Baby goats are called kids.

A male goat has hair
on its chin. This is called
a beard.

Goats eat plants
and grasses.

Goats have soft hair.
It is called wool.

15

It is used
to make clothes.

One kind of goat is
an Angora goat.
It has curly hair.

Farmers get milk
from goats.

This milk is used
to make cheese!

Words to Know

beard

cheese

wool

Index

24